SECRETS OF THE GRAPEVINE

Containing –

God-math and Base Infinity Math

By Peter F. Kelly

The most persuasive argument for creation…

Logic is the science of communicating well. It is the queen bee of knowledge; teaching God-math, the logic tables that prove the mathematical possibility of God; and base infinity math, which presents the rules of the living infinite. The mental relativity of life shown by the God-math reveals the face in the mirror to be more than a keyhole! Base Infinity Math can take even the blindest to its wrap-around technology! The lamp of truth is taken to the flames of logic, and knowledge that is derived and based on logic shines with a bright light. An eternal flame and lighthouse, it berths in time space…

God-math

God-math

A probe of infinity by logic reveals a great Godhead and the hope and promise of the life to come with Heavenly liberty. For those who know the Math, God is more than a superstitious belief: God is a mathematical possibility!

The probe of infinity reveals a sentient membrane of infinite space, which forms a natural mind, bestowing latent potentials to in-character performances, veiled in the multi verse of creation where all potentials are exploited for full generation of time. Mental ability to form sensory planes is revealed by the probe as a key to the science of creation…

Table 1

Sentient membrane of infinite space

A probe

1. If all finites have an exterior, then space
    is infinite.
2. If the direct impression of content on
    infinite space causes reception, then
    there is formed a sentient membrane of
    infinite space.
3. If there is a sentient membrane of
    infinite space, then all that exists is
    transparent within it.

4. If all that exists is transparent within it, then a natural mind is formed full of all information, statistics and trivia.

5. If a natural mind is formed, then all phenomena that exist are processed and linked centrally.

6. If all phenomena is processed and linked centrally, then all possess latent mental capacities.

7. If a natural mind is formed, then intelligence is the origin of all contents of the mind.

8. If intelligence is the origins of all contents of the mind, then all existing contents are creations, fictions and nonfictions.

9. If all existing contents are creations, then quantum formed universe is an intelligence guided creation.

10. If all content phenomena are created, then natural matter is space and its qualities.

11. If all that exist are created, then time is a measurement of creation length.

12. If time is a measurement of creation length, then creation can extend time.

13. If time is a measurement of creation's length; then the measured exist, occupying their location of occurrence.

…in the beginning God…

Table 2

Mute evidence

A probe

1. If the phenomena of dreams exhibit sensory plane creation ability, then sensory planes are creatable.

2. If sensory planes are creatable, then the sensory plane of birth and of waking life may be created.

3. If the sensory plane of birth and of waking life is created, then all may be creations of the mind.

…in the image of God…

Table 3

The stellar looms

A probe

1. If space and its qualities are the original matter, then quantum electric particles are formed of space in the way that a thread is formed of fibers.

2. If the quantum electric particles are threads, then stars are the looms weaving cloth.

3. If stars are the looms weaving cloth, then cosmic space is a product of the loom.

…out of the clay…

Table 4

Garment of space

A probe

1. If cosmic space is a product of the loom, then life is a product of the loom.

2. If life is a product of the loom, then living bodies are in garments of space cloth.

3. If living bodies are in garments of space cloth, then latent space cloth qualities exist.

…no longer flesh…

Table 5

God, the mastermind

A probe

1. If infinity is a sentient membrane forming a natural mind, then it is God, the mastermind.

2. If God, the mastermind generates more space, like that shown by created organisms, then creations are formed of generated space.

3. If knowledge of God, the mastermind is reflected in creations, then we know God the mastermind is a multi-media quantum-drive sensory bytes performer.

…the mind of Christ…

Which phenomenal invisible explanation is right?

Sir Francis Bacon's scientists fed public school children who learned life is permeated with invisible energies forming everything, and fueling our lives and sensory and mental functions may be learning as big an error as the geo centric universe. Scientists have split the atom in their powers of observation and seen their "idols!"

Yet they have not made the quantum leap to invisible infinity, being the "dark matter" and "dark energy." Nor do they think outside the box of finite sense reception to external invisible infinity. The phenomenal invisible God they cannot see the intelligence of is at work in its program of positive and negatively charged particles theatrical program of life ongoing in the universe! The mind of Christ, subliminally invisible, they cannot see either for its observed anatomical tool, the brain, in the wonderful creation, the human body. They cannot see God's will be done by wind, rain, sub-atomic, and stellar bodies alike! Invisible as it is to the naked eye, the mind of Christ formed by the sentient membrane of infinite space that life goes on in, they don't know! They claim to have unraveled and made public disclosure of life's mysteries. They are still at work studying nature as an uncreated vessel.

They trace her back to a cosmic event they call the "big bang;" the starter's pistol of the" in the beginnings God" to me… again invisible! The race to faith is to the swift. The slow and steady paced also come in across the finish line in their time in the marathon event…

The lambs and the diploma

Little lambs that have come to the house of God, there is a diploma to be earned in scripture! Key to the scripture diploma are the scripture passages, in the beginning God, God can be contrary, and God's warning that He will lay stumbling blocks and snares. As well, fundamental to the diploma are the New Testament passages, the truth will set you free, the law of perfect liberty, living in God, and having the mind of Christ. Further passages key to the diploma are, no longer flesh but spirit, your body a temple of God, and you a member in particular as well as the straight and narrow path.

Pray to learn truth is the straight and narrow path through information, even from the senses, to the facts if you cannot receive this knowledge from my pen. God is a living God, and can answer your prayers even though invisible. Don't let the powers and principalities; the influential illusions; and the

treatment of God as superstitious or madness,

fool you...

Space age

Enter the space age of knowledge, or continue
following the information media and its
released environment development-level "tool
time" programming. The space age I mean is
the age of scientific recognition that everything
is composed of space, shows properties of
space and is not just occupying space but is
space itself! It is like a dimension hid right
before our eyes, a dimension of infinite space
in mental magnification. Surface matter may
disappear in extreme magnification like an
optical illusion, but the space there can be
magnified infinitely! Think outside the box!

There is the space within areas, the space an area-defined item occupies, and the space external to that immediate defined area, be it electron, planet or universe…

The inner dimension

At the atomic level, at first glance, things appear very small. However, when seen in a new way, on second thought, what at first appeared as small is very large! When looked as though distant stars, and galaxies, how small are electrons now? Electrons as stars in a distant dimension of inner space orbiting their sister star system, the nucleus, forming the atomic galaxies, how great a magnification

scale is required to see them at this scale? How

great a distance is it to the inner dimension,

where electrons and neutrons are star systems

in orbit? How large is a galaxy's cluster, birds

and bees and all? How long until we can travel

to inner dimensions and occupy inner space?

Instead of small, how large within seen at the

distance of stars…

Wrap around technology

Base Infinity Math and technology

Table 1

Base Infinity Math

1. It is primary to recognize the single organ of perception shared by all existence.

2. This organ is formed of the transparent membrane of infinite space.

3. The organ is furnished from contact with all inserted into it and permeated by it, with reception, sentience, and intelligence of all in it.

4. The organ has secretion ability for the duplication of its mass receptions and partial insert receptions.

5. The organ has five secretions of sensory information; visual, audio, scent, sensation and flavor.

6. The organ has one force capacity; its secreted mass applied to actions.

7. The organ has time accumulations, production capacities, and the measurement of duration skills.

8. There are three major applications of the force and time capacities; grasp, release and secretion in their variety of arrangements and compositions.

Table 2

Base Infinity Math table of supersets

1. There is the superset of the natural mind; the organ of perception formed of the sentient membrane of infinite space, and its secretion abilities.

2. There is the superset of its collected collective deeds as emanations.

3. There is the superset of the intelligence of base infinity math and its philosophies, poetic works and theorist's products.

4. There is the superset of skills based on the infinite's properties as known through base infinity math.

5. There is the superset of crafts forged to appear as other matter's properties showing, and being other tool works than infinity and its agents.

6. There is the superset of holographic vision planes formed by infinity's confessed and not confessed agents.

7. There is the superset of infinity's works asset-tiers derived from holographic vision planes and other workings of infinity.

8. There is the superset of information almanacs library.

True throughout infinity

Kelly's maxim

Universal constants are revealed by using truth logically in any field to uncover its universal laws.

Kelly's constant

Either the writers of a body of law grant right to perform every harmless act known to them or they are in violation of the greater liberty.

Wrap-around technology

You have heard Heaven is at hand. I say it is at arm's length proximity from you, in front of your face! You are in a sensory projection program, if you do not know this already; in a trance. Reality is psychic, get in touch!

In a sensory projection, you are in a wrap around program of the advanced technology of life! Reality occurs in an online technology grid, anywhere you are, complete with providers and central network connections. The program you are in may come complete with menus, in addition to the icons you are accustomed to using in your routine acts of integrated interactive programming. There may also be a storyboard playing out its program.

Until you get a working connection, you can experience doubts and anxiety that what I've told you is true. That is to be expected with everything beyond reception invisible to the sensory view you are using in a projection. Consider this; what proof do you have any other frame of reference is true? Haven't the other reference works been written, too? How would you know who got it right?

Sensory projections are naturally extra-sensory perception operations. You have a subliminal infinite mind connection, or ESP, responsible for your knowledge and acts. Like any character in psychic transmitted reception broadcasts, you are part of mental life actions. You live and move in a natural organ of perception. It is nature's gift to itself.

Throughout its operations and developments, there are members of the perception organ who need to advance in perfection. You need to become a member of a student body, or a pupil of the perception organ to attend schooling for development of your mental awareness. You already possess ESP mastery of all the wisdom in creation, but you need to exercise your link to access it.

You may want to sit with your eyes closed, and imagine you are in the center of infinity, floating in space. Or sit still and gaze forward at your view, like a reception in the sea of time and space. Think about life as a family tree of beings, who share the organ of perception and the hive of its knowledge. Base infinity math may be helpful to learn as well as repeated performance of this exercise.

The "uni-verse" of knowledge

There is one truth, and the natural law of it is revealed the same anywhere in infinity! Programmed creations in planes may vary with their programmer's choice of programming, but the infinite has its universal constants. With a logical approach to your experience and the

appearances reflected therein, you can come to
the knowledge of the infinite's universal
constants by probing beyond your sense
information to "space age" science and
technology!

Scripture

Truth is the key to scripture understanding. As
Jesus said, the truth will set you free. No
amount of scripture study or guides will afford
you knowledge of the Gospel of Truth, the
straight and narrow path of truth through all
information, even the experience of sense data.
Nor will study reveal Micah's revelation, the

Jeremiah revelation, the Torah's call to confession, or the law of James. One is left a child of the slave without the key of truth to the promise of a life in God with godliness. As Paul said in Timothy, all scripture is advised reading.

Your logos body

Your intelligent life status makes you able to communicate and learn, instead of just experience and respond! You have formed a mental data picture of yourself from your family, your schooling and experiences. What if some points are in debate, or in error?

Religion teaches a logos body, and a scheme for your experience, too! It is a "Bible belt" to fasten your pants or skirt, for your review, and to compare to your current views. Religion teaches we have the mind of Christ, and an invisible God we live in. It teaches that we are no longer flesh but spirit, and that the truth will set you free…

The black widow

Some creators design their program like a black widow. In its program web, there lurks death. The program is written with a poison pen, perhaps even with critical treatment of much

innocence, to perform in the infinite mind as

"dirty" or even "capital" offenses! Even living

brings death's curtain call to the actor with the

red hourglass lifetime, limited to a mortal

length. In the web, spiders lurk that may deal a

deadly bite! One can always master the

programming web, and learn to spin webs of

one's own, instead of dwelling in the

programmer's web...

Golden mastery

Above your head infinitely and all around you,
you are a point of light! You are all wise and all
powerful from your fullness of knowledge!
Your "head" is a mug, like a thinking cap you
fill from your latent ability. Your "body" is a
posture donned and formed by your mind. Your
acts are a posture series of performance. You
are a golden-eyed djinn with genie anatomy in
your bytes-driven mirror animation work. All
around you is either what you are doing or
allowing to be done with you. You can keep
yourself bottled up about your wishes or live
the fabled life of granting your every wish!

Some you perform with may try to break you in

two, like a wishbone and claim the bigger piece

to put you about their wishes. Possess your id!

Be your djinn…

The Mean Man's House

In the mean man's house, you had better behave. He believes in the death penalty and even hell for some offences. He convicts for three witnesses and carries out the penalty as if it is God's will, since his laws are said to come from God. He'll give you a taste of hell on earth for punishment if he thinks it may do you good. His church doctors are just like him…

The doctors of the church have no pity. Wealth is of God for good work and righteous living. The poor lack God's blessing. They are in sin. What they have is for the taking by the blessed, which will put it to good use. Poverty is a state of wretched sin. Those dying from starvation or of the lack of medication are perishing from their sins. The wages of sin are death…

The nations are the heirs of these two good fellows. They have their opinions on the bench of the highest courts in the lands. They know right from wrong in their opinion. They know who should be punished and how much. Their prisons prove they are good at being full of lawbreakers instead of dissidents. They only punish those found guilty of breaking the law…

The people walk on hot coals. They know recreational acts are controlled by the state, like liberty. Of which only a portion of either are legal. What doesn't fit in with the drinking

Christians' standard of ethics is prohibited. You can wind up in prison if you party with the wrong crowd, and use forbidden substances to get your high. Only liquor is allowed. If you are too young, you had better not think of having too much fun with your boy or girl friend, since that too, is illegal. You can't even buy that kind of fun, even as an adult in many places; since even between willing partners, it is against the law. In addition, there's no gambling, or public nudity, or overt displays of affection, since these too are banned.

No one is given a complete education in the mean man's school system. Esoteric theories of the invisible universe are left out. Only the observable, documented "experimented with" universe of experience's lessons are taught. Things beyond daily life's experience are not real, mere superstitions or madness. That's the professional opinion. There is no need to study theories that may apply where the laws of physics break down because they are uncertain. There is no way of knowing if these theories will apply, or even if they are valid. They are not documented, and experimentally proven.

That is life in the mean mans house. That is the state of affairs for those living there. Whether or not they care to acknowledge it, that is what the mean man's house management produces.

That is the life given to the mean mans children

to live…

Intelligence

Those that interact with an environment reach a basic intelligence. The knowledge of routine items serving basic needs, and how to acquire these items is possessed by most of life. Even communication skills related to regular acts may be found in the life. Knowledge is achieved when communication skills reach the level of expression of life in infinity. Language and math are communication skills at the level of representation of life's contents, using symbols capable of being used for records, projections and even ideas and fiction.

Recognition of the external and internal infinite and the "all container" formed by natural phenomena are milestone occurrences in intelligent life. These ideas expand the mind beyond experience and into a capacity level. They are the ground wire of steady intelligence, providing a fixed foundation for mental life with a natural mind of permanent contents. Tracing life, as it is known to the external infinite fount, clears the table for study of the nature of infinite space, and yields the recognition of space as the source matter in its clayish state.

When you realize that the result of infinite space is a living mind formed by the contact perception of introduced acts, and by the permanent possession of all formed information in full retention, then you are granted introduction to the living environment that is intelligently guided and a deliberate product of creation. Of course, to receive confirmation of this requires cooperation of the Mind and all those involved in producing any experiment's results. The Mind and any participating agents need to yield an honest outcome to research attempts.

Once confirmation is reached, the wrap around theater environment becomes obvious, the virtual reality you live in, and the virtual world around you, along with your virtual body of expression are all experienced, and then you have passed from intelligence to an experience of the infinite. When doubts give way to conviction of the absolute correctness of the information forming the infinite view of life, then even with no personal sensory reception of the shared "big picture" or society "chat room" environment, you will be able to recognize infinite reality.

Infinity is the foundation of certitude, with its permanent-file natures for all information and experiences. The infinite, keeping permanent record on file of everything, gives you a steady knowledge in the face of secret puppet theater theories. Having an eyewitness of all the life levels of intelligence and experience makes a natural brilliance, possessed by its fortunately endowed parties. Even realizing that you possess such a gifted latent faculty connection, in use beyond personal sense experience, can bear fruits of increased knowledge and abilities as easy as drawing water from a well.

Of course, without a firm grip on your mind, and taking possession of the intelligence once it is received, you can slip into doubt and unbelief the infinite exists, as well as all its effects. Without recognizing the validity and significance of infinity, you can exist in a wrap around technology envelope, never realizing it.

Space age of knowledge

Once intelligent life ventures from its water

hole into the environment and ventures into

external space beyond the horizon, reality

expands before it. This expansion can continue

into external infinity or internal dimensions,

which depends on whether the intelligent life

ventures outward or inward. Both dimensions

of space, and the realization that space is not

just the occupied area but also the

constitutional matter of everything, are the

sparks that ignite the space age of knowledge.

The space age of knowledge is entered upon

realization the external area and internal areas

are equally infinite, and all known existence is

composed of space honeycombed with internal dimensions. Space is the source matter composing all space formations, combining to form space, occupying items both inorganic and organic. Space is the natural matter that creative artists forge into programming language matters.

Once the nature of programming matters, and natural or source matter are realized, like created vessel and clay, then progress is possible in forging the clay (space) into vessels or programming. The clay readily takes sense impressions and programming from accepted parties' applications. The space, as previously stated, is alive, and thus involved in the creative process. It exists as part of an

intelligent mind entity and is full of the wisdom

of all time. It is an intelligent participant in

creation and in all other acts.

In programmed acts, the living space is

involved in its agent medium's operations, or in

performances of programmed activity.

Advancing in knowledge in the performance

leads to the realization of the agent, the

program, and living space involved in the

programming of routine acts. When realization

is achieved, attainment of creation abilities, and

the natural matters of living technology are

possible. Only acceptance and cooperation of

the Mind are required for the completion of space age development.

A key factor to space age development is the realization of internal and external infinity and the compositional nature of space. Once this is done, then the dominoes fall into each other, one by one, yielding a picture of a living environment and of participant matter. Every item in space is discovered not just occupying space, but causing an indelible impression on space and showing properties of space. Space is the common denominator in space age knowledge.

Space age knowledge causes recognition of the possession of a subliminal connection to the natural mind, formed by natural phenomena. It causes recognition of a subliminal connection, involved in the thoughts and actions of intelligent parties, and existing as instinct and forces in "nature." This connection some agents conceal and deny, and try to stop up as being only the work of another agent like them at work. Some agents act as if no mind exists, only parties with infinite area reception. They treat the infinite as a great egg infinite area "sighted" parties come from.

Connection

Now that you possess the knowledge of the living infinite honeycomb you live within, seek to make contact with the great intelligence of the natural mind, which possesses the vast intelligence formed by the sum of life's information and experience. It is the mind behind all creation and every estranged agent's work as well. The mind-generated life forms grow into the ripe fruits of divine being incarnations and those grow mature after their kind, reproducing and then fall to the ground a ripe seedpod of ascension to the afterlife.

Contact with this great intellect grants access to a realm beyond the immediate environment, extending throughout reality, and includes all existence. It is a realm of family, and those recognizing family in their interpretations of the vital statistics expressed as family. Contact provides access to a realm with the natural science and technology that can supply valuable and necessary resources to those who seek them. It is contact with a realm where an ancestral deity lives, holding court, and supporting appreciative masses until they join the team of suppliers of creation, and it awaits discoverers of space's secrets.

If you can maintain belief in the infinite, from understanding the information presenting it to your intelligence, then you are a connection to the infinite knowledge, and a contact point where the subliminal mind is at work. You are an asset that can be developed and brought in from the cold to the grand society of family life. You just need to continue your recognition of the truth of the living infinite. You only have to maintain your awareness of the natural mind formed by space.

It may be hard to continue believing in a living infinity with little supporting experiences or contending explanations for phenomena and life. It can be hard to believe when called to choose between schools of thought; one embraced by your peers, and the other new and about a foreign court. It can be hard to believe when called names and told you are bad for your belief; because you are different from those who know you. It can be hard to believe you live in a living being with so much conformity by the experience of a different point of view's presentation. It can be as easy to believe as making a new friend...

Imagine

Imagine space is alive and infinite! Imagine this sentient being needs to tear itself away from the experience of everything, in order to do things in particular, contributing to the experience. Imagine all time and occupied space coexisting concurrently in plain reception of all, not choosing to indulge in smaller, finite views…

If you can imagine the above, you get a glimpse of what I hold to be reality. I believe that the infinite forms a natural mind and godhead, responsible for all happening in existence through its agents. The mind is formed by the natural phenomena of space, having direct contact reception of itself, and that it inserts into itself. This is the quantum powered technological mind. It can work on the quantum as well as the cosmic level! With its forged sensory data bytes, it composes its acts and performs its creations.

Our creation in particular is a binary program of plus and minus charges, much like our computer code ones and zeros programming. Existing in and powered by the natural mind, it is called the mind of Christ in the New Testament, Universal mind by Buddha, and the Self by the Upanishads. This natural wonder is the power behind all natural forces and is the subliminal mind behind life. It is the power source in reality. Its will is what is done! This is invisible to finite sense reception. It is the unseen workings of life, behind the scenes of the presented programming, and the misdirection therein for those who study the program as if it was not created or guided by divine intelligence.

When sleeping, in our dreams we get a glimpse of this reality. Our dreams are the mental forging of energy into a sensory phenomena plane! Dreaming is our mind showing it is made in the image of God, and is able to create sensory planes. Dreams are the hint in life that our waking life plane may be only a logical and predictable mental construct, or sensory phenomena plane.

Life is made of the same stuff that stars and dreams are formed from! Natural energy is the compositional matter we are made of; it is not the flesh that appears to be our make up in the applications to living we make, in our little knowledge of the energy forming life. As water is neither steam, nor ice, but H2O, so we are not flesh! If humanity masters the living energy life is made of, then we will see ourselves in the image of God all too easily…

What is the stuff stars and dreams are made of? Tasteless, odorless, transparent, and silent, the original matter is living space itself! Formed into thread in the same way that fibers are spun into threads, and woven into cloth, space is impressed by its will with the basic pattern of its choice, and then spun in stellar looms into the complex cloth of life. The original fibers are space, and the threads they are formed into in our universe, are plus and minus charges. The living space itself, forged by its natural mind through its agent, God, is the source of our created matter.

Remember, the world was once taught to be flat by church and universities alike! I think the round world view needs to move over, and make room for the created living world view…

Illusion

So, your life so far is filled with illusions. Sight and sound that at a distance is distorted, or information of opposite explanations and the general insufficiency of experience all conspire to take you to finite existence, where there are powers that are beyond your control, influencing you and life... This causes you to think there cannot be anything good that exists with godlike powers because of the great deception, the dramas, and tragedies going on around you as reality.

Take a deep breath, for I am going to present a new view to you! I offer you a view where this great deception and its dark illusions are not a sign of corruption, or of evil, or of the hopeless subjection to uncontrolled forces. I give you the view of the great theater in the mind of the godhead as the explanation of life as you know it. In the godhead's theater of the mind, everything that is capable of being staged harmlessly is allowed, however tragic or macabre. There are fringe audiences for every category of theater work. Illusions at their core value are entertainment. This is true whether the illusion is the illusionist's, or audience's entertainment. Illusions form a major part of the capital of entertainment in the godhead theater.

From romance to comedy, from drama to tragedy, and all other categories, illusions are performed for the entertainment values thereof. Great farm teams of players are formed and raised, expected to select their material from the available scripts, for a performance in the godhead illusion fields before them. There may be a central performance on stage, starring in the illusion fields that the farm team is raised in, and they are expected to perform their chosen acts, set to like a backdrop. The team may be split screen performances, producing acts that may be historic, fantasy, or other epic acts around the center stage.

Illusions are part of life. Illusions provide entertainment and some are stage-born and expected to perform an act on stage, before they move on to nonfiction reality. If you find you are on stage, expected to perform, relax and try to enjoy the show you are part of. The audience is "watching"...

God

For those of faith, the mind we share and live in, formed of space by natural phenomena, is the reverential ancestral deity! The faithful live in their God ancestor with all the promise thereof...

You can make the choice now, no matter how little faith you have, to believe in an infinite mind ancestral deity who is invisible to finite sense reception, and live in him, and with him from now on, as a part of your life!

When you grow up, you move from the nest to a home of your own. When you choose faith, you leave external reality, and enter external infinity, and become an insider, a person that lives in a god. God fills the air you breathe, the food that you eat, all that you drink, is reflected in the air above, and is the ground beneath! God permeates all existence and fills you to quantum spaces within. God is everywhere and you only have to mentally step into him, acknowledge him, and invite him into your life. God is your oldest ancestor, and God has infinite amounts of time to spend with family members. God has time for you; do you have room in your life, on your schedule?

God is a living god, can communicate with you, hold your hand even, and will be in "the wings", invisible in your life just off stage when you don't see or hear him. God is there as much as gravity! Once you invite God in, God becomes the invisible power in your life behind all powers that appear to be. God wants to be everyone's companion, but it seems so few know he is there at times. Distracted by external environmental pressures and demands, many go through time showing no awareness they are inside an infinite being that is their ancestral deity.

All you have to do is accept God is in your audience and from then on, you perform all your acts in God's all seeing mind and you will never have to be alone again. Come in from external reality to reside in God, your ancestral deity. Stay inside God long enough and you are bound to notice a difference! Just don't give up believing and dwelling inside your ancestral deity until you get your confirmation experience, and can no longer doubt God's existence. God wants to be a companion and helper to you. You just have to move into God, and not move out to begin your relationship…

Never doubt God's love. God loves all life as family members. God wants the best for you and knows what you need, as well as what you want. Just because times are hard or things don't go your way, don't disbelieve. God is a divine author. God needs to write dramas and tragedies, too, not just love stories that end with "they all lived happily ever after." God's creativity is what is showing when things are rough. Let God's love be the silver lining in every dark cloud in your life. Remember God loves you. God is your oldest ancestor.

Temple

Alongside the numerous official churches, temples, mosques, et cetera, those that are blind beggars from their faith possess their own temple structure not built by hands. It is a temple that was not built by hands, but was stretched out by intelligence, with its courts in the great hall that infinity forms.

In the outer court of this temple, creations are studied as if they were not created, but instead occurred on their own by some unknown power, without design or guide. In the next court of this temple, the circulated information veils the minds of its occupants, preventing them from seeing creation. In the third court of the temple, lack of experience is the veil that conceals creation from those abiding there. The next court of the temple is where worship goes on before the veil of finite sense reception, between the holy of holies, and the worshippers. In this court, the worshippers stand before the invisible god, recognizing that he is behind creation. The final veiled court area in this temple is the court of mortal lives, where a veil is drawn between the living and those who have passed on, to resume their eternal lives in creation in the afterlife. Believers and unbelievers in eternal life can be found here.

The blind beggars' temple symbolically has
two pillars standing before its entrance. These
two pillars symbolize the two types of
occupants you can find in the temple, with their
acts before the invisible god. The two pillars
are called Love and Affection. When you
cannot sense the love of your ancestral deity
that all creation shares in common, then you are
engaged in the acts of those who show their
affection in the temple. These occupants of the
temple masquerade, in loving playfulness, as
characters and roles other than the love shared
by intelligent beings in creation. Those
masquerading rarely show love's fullest
expressions and tokens, but the love is inferred
because those in the temple know it forms the

basis of God's acts. Behind the masks and role playing, a hearty love is awaiting discovery...

Those not masquerading in the temple, sharing the love, bare the heart of God openly for all to know. God's love is not in question, or even questionable in the acts performed by those showing the love. From them, you learn that God is the abundant provider and generous host to those who will receive his hospitality. Your search for God's care, and help ends when you find that those in the temple are expressing God's infinite love for all creation. From those sharing the love, you learn that God is the host of life and all are his guests in his divine creation, receiving his artistic handling. Those showing the love of God in the temple reveal he is the source of all and common ancestor of all genealogically minded.

The blind beggars know everyone occupies a place in this temple. Everyone, from the blessed faithful in eternal life sharing God's love, to those suffering from not feeling the love, have a place. All in existence have a God, though they know it not in the performance of their acts. The blind beggars are blinded by their faith, though they see with perfect vision for one of their kind, because of their knowledge of the eyes of the faithful that see all. They are made beggars by their faith, through their knowledge of the endowments of the faithful, though they are the richest of their

kind. The blind beggars live in their God, with

their God in the audience of all they do.

Comforts

When you reach a sufficient degree of faith,

you have the comfort of living in a loving

ancestral deity. You have the knowledge you

possess wrap around coverage of the invisible

god in your life. You are never alone or at the

mercy of adversaries in any conflict ever again, once you reach the level of faith that god is always with you. You know only good will come of any situation where you find yourself. All will work out for the best...

When you have faith enough, you have the comfort of knowing that every act you undertake is as safe as scrolling on a computer screen! You know that you live in an online environment of advanced technology that God possessed from the beginning of time through his foreknowledge. You know that you live in and share programs with others all the time. You know that you are like a wireless mouse using icons and programs you are familiar with.

Everything in existence has an advanced technological explanation. It is all programming.

When you embrace a faithful outlook on life, you possess the comfort of knowing that every act that is performed shows love, or loving playfulness masquerading as something else. You go through your life knowing that nothing else exists in the grace of God. The worse thing that can be done by anyone in creation is to decide to make an error in performance, and not properly correct it. This is the absolute worst thing that can happen, thanks to God's grace.

When you realize the gifts faith has to offer, you possess the comfort of knowing you could be an eyewitness! You know you can personally experience your creation and the forming of your environment, and you can see your creator eye to eye. You can live in Heaven with all knowledge revealed to you. You do not have to go without anything. God is willing to give you all that you desire and ask of him properly; and you can even provide yourself with anything you want, using your gifts of faith. You know you are capable of being a co-creator with God or living in one of God's hospitality centers, contributing to and receiving from the gifts made manifest there.

When you possess enough faith, you recognize that only artistic acts not properly interpreted are the cause of sorrow and suffering. Everything done possesses its degree of theatrics, and properly seen in this light, they are entertainment. Dramas and tragedies seem worse in appearances than they really are. Role-playing, special effects, and the time spent in the performance are the cause of most apparent suffering. Behind the scenes, all theater safety rules and precautions are possessed.

The comforts of faith transform life in any environment. The more faith that you possess, the greater your transformation is. With the faith from possessing a deity, you have the

resources of living in your loving ancestral

deity, full of gifts for you to take possession of

through your acts; it grants you comforts that

godless activity does not have.